The Operation of Infrastructure

I0427070

The structure that changes our world

Koso Brown

Contents

Introduction

A nation's economic growth and development are closely related to its infrastructure. It provides access to necessities like health care, education, food resources, transportation, employment opportunities, and more, and it catalyzes the creation of solutions that reduce poverty. Additionally, it raises production and raises living standards in a lot of places. Societies are supported by institutional and physical infrastructures. "Infra" is a Latin prefix meaning "below" that appears in the word "infrastructure." Modern, industrialized living would not be feasible without these fundamental institutions. You will gain an understanding of the complexity and variety of infrastructure as well as why it is necessary for day-to-day functioning in this reading.

Chapter 1

What Is Infrastructure?

The economic, social, and physical frameworks that uphold society are referred to as infrastructure. The construction of infrastructure is essential to the efficient operation of a contemporary, industrialized country. Infrastructure falls into two main categories: hard and soft. Hard infrastructure refers to the actual structures that sustain daily life, such as electricity grids, roads, bridges, and highway systems, as well as the products that keep them running, like buses, trains, and mass transportation. Human capital, or social and economic elements including banking, telecommunications, and healthcare services, is referred to as soft infrastructure.

What Are Infrastructure Projects?

Public or private attempts to establish, maintain, or improve infrastructure are known as infrastructure projects. Large-scale infrastructure, which includes highways, bridges, and roads, is typically financed by taxes as part of the public sector. Public works projects generally entail collaboration between local governments and private sector organizations to provide public infrastructure, including parks, schools, hospitals, and law enforcement facilities

Understanding Hard and Soft Infrastructure

Infrastructure is commonly classified as either soft or hard. The actual, physical construction of things like highways, railroads, bridges, and tunnels is known as hard infrastructure. Soft infrastructure is the services required to maintain the economic, health, and social demands of a population.

Hard Infrastructure

Hard infrastructure is the tangible framework required to operate in a contemporary, industrialized country. Roads, highways, bridges, and the equipment needed to operate them, like cars, oil refineries, and transit buses, are a few examples. Hard infrastructure includes technical systems that are essential to the running of businesses, like networking gear and cabling.

14 million people work in industries directly tied to infrastructure, according to the Brookings Institute. Nearly 11% of all occupations in the country are in the infrastructure sector, including those of locomotive engineers, electrical power line installers, truck drivers, and construction workers.

Soft Infrastructure

To sustain an economy that provides the public with services like healthcare, banking, government agencies, law enforcement, and education, soft infrastructure refers to the human capital and institutions that are required.

Soft infrastructure investments focus on how people live well and engage in daily activities. Proposals for soft infrastructure, such as Medicare expansions and community college with no tuition, were the focus of President Biden's Build Back Better Plan in 2021.

6 Kinds of Facilities

Here are 6 different examples of hard infrastructure examined in further detail.

1. **Water infrastructure:** Maintaining human existence requires a steady supply of pure drinking water. Water systems supply electricity and water for irrigation and drinking. Water pipes, wells, dams, gutters, and pumping stations are examples of the infrastructure related to water delivery.

2. **Waste management:** Waste management is gathering and handling waste. Trash

collection, recycling centers, sewer systems, and landfills are a few types of waste-management infrastructure.

3.

4. **Energy infrastructure:** The grid's power is provided by nuclear, natural gas, and coal-fired power stations. By making investments in renewable energy sources like solar, wind, and geothermal infrastructure, innovations in power and energy infrastructure aim to promote environmental sustainability.

5. **IT infrastructure:** The interchange of data and other digital resources is supported by information technology infrastructure. Data centers, cloud computing platforms, and operating systems are a few examples of IT infrastructure.

6. **Infrastructure for transportation:** The physical systems that facilitate travel include roads, highways, toll roads, railroads, airports, and air traffic control. Public transportation,

such as buses, trains, and subways, is also included. With the investment in electric vehicle charging stations and the limitation of greenhouse gas emissions, newer infrastructure regulations may seek to mitigate the effects of climate change and establish stronger environmental protections.

7. **Telecommunications:** Broadband access and internet connectivity are examples of the technological resources that make up a telecommunications infrastructure. Radio broadcasting, satellites, mobile network systems, and phone lines are all included.

Maintaining Infrastructure

The ownership of infrastructure typically determines how it is funded and maintained. A large portion of the infrastructure for public education, water, and transportation is owned by the government. While some infrastructure may be fully owned by private entities, the majority of infrastructure is held by state and local governments and is frequently subsidized in part by the federal government.

In addition, there are public-private partnerships for infrastructure upkeep. To run and maintain the Chicago Skyway Bridge, Cintra and the City of Chicago signed a 99-year lease in 2004. The agreement states that Cintra will get all toll and concession money from the bridge, with the city receiving a $1.82 billion cash infusion and getting out of the maintenance business.

The bipartisan Infrastructure Investment and Jobs Act of 2021 created a new government subsidy that will allow AT&T, Comcast, Verizon, Spectrum, and sixteen other providers to offer high-speed internet plans of at least 100 megabits per second to eligible low-income households "for no more than $30 per month" (White House, 2022).

Several infrastructure initiatives have been launched by the US, such as the $305 billion transportation infrastructure bill in 2015 and the American Recovery and Reinvestment Act of 2009. The $1.2 trillion Infrastructure Investment and Jobs Act, signed by President Joe Biden on November 15, 2021, will finance the reconstruction of roads, bridges, water infrastructure, internet, and other infrastructure.

New incentives and investments in infrastructure development are also included in the 2021 package. For example, $7.5 billion would be used to encourage electric automobiles, and $65 billion would be made to guarantee that every American has secure, high-speed internet access.

Chapter 2

Understanding Infrastructure

Originating in French, the term infrastructure was first used in the late 1880s. Infra- means underneath, and structure means building. An economy's infrastructure can be thought of as its structural basis.

A city, state, or nation's electrical grid is one example of an infrastructure system or structure that requires physical components. In addition to facilitating residents' participation in the social and economic life of their community and supplying them with basic requirements like food and water, infrastructure also includes buildings, machinery, and other comparable physical assets like roads and bridges.

Public financing, control, supervision, or regulation of infrastructure is commonplace since it frequently entails the production of public goods or goods that lend themselves to production. Usually, the government produces these directly or they are produced by a legally recognized, strictly regulated organization. Constructed at Virginia Beach in 1789, the Cape Henry Lighthouse marked the inception of federal funding for infrastructure projects.

Private businesses occasionally decide to fund the development of a nation's infrastructure as a means of expanding their operations. An energy firm, for instance, might construct railroads and pipelines in a nation where it wishes to refine petroleum; both the nation and the company would profit from this investment.

It is also possible for individuals to decide to provide funding for certain public infrastructure upgrades. One could, for instance, donate money to support local police enforcement initiatives, hospitals, or schools.

Characteristics of Infrastructure

The following are the fundamental qualities of infrastructure:

1. Infrastructure supports the main industries that produce goods and services, including agriculture and trade, both domestically and internationally.

2. Among the infrastructure services are:
   - ✓ a financial system made up of banks, insurers, and other institutions
   - ✓ hygienic system with access to clean drinking water
   - ✓ roads, trains, ports, airports, pipelines for gas and oil, communication infrastructure, power plants, etc.
   - ✓ Health system including hospitals.

3. Certain services have an immediate impact on the production system's operation, while others have an indirect positive impact on the economy by supporting the social sector.

Chapter 3

Why Is Infrastructure Important?

Infrastructure is vital to society for several important reasons:

- **minimizes supply chain interruptions**

Supply chain interruptions can have catastrophic consequences for an area's economy. The goal of critical infrastructure is to minimize disruptions to economic activity, importing and exporting, and the labor force.

- **Development of the Economy**

There is a relationship between infrastructure development and economic development. Infrastructure boosts a nation's productivity as well as the standard of living for its citizens, which both aid in the economic growth of the nation.

- ✓ Agriculture depends on the construction and expansion of irrigation infrastructure.
- ✓ Industries cannot develop without the expansion of power and electrical generation, transportation, and communication.

An economy's national income and infrastructure development are positively correlated. Improved quality of life, more productivity, and better levels of revenue or output are all made possible by infrastructure, which supports economies.

- **Improved Standard of Living**

Well-developed infrastructure leads to improved quality of living.

- ✓ Improvements in sanitation and water supply result in a significant reduction in both the intensity of the illness and the morbidity from major waterborne diseases.

✓ The condition of the networks for communication and transportation affects access to healthcare. However, particularly in densely populated places, air pollution and safety issues associated with mobility may affect morbidity.

- **Facilitates the Outsourcing Process**

Outsourcing can be advantageous for a country with developed infrastructure. India is growing in popularity as a site for call centers, BPOs, KPOs, and other related industries because of its robust infrastructure and IT support system.

- **Produces Employment Opportunities**

Job creation is influenced by infrastructure. The construction and maintenance of roads, power plants, electricity, etc., employs a large number of people. A strong infrastructure enables a large number of people to work in trade and industry.

- **Productivity Is Enhanced by Infrastructure**

Health and education facilities are part of the social infrastructure. These institutions offer education, healthcare, and skill development—all of which are prerequisites for increasing productivity. This suggests that productivity will rise as a result of increased efficiency. Furthermore, this leads to an acceleration of the growth process.

- **The Infrastructure Promotes Investment**:

Infrastructure stimulates investment. An efficient transportation network, for example, would surely stimulate investment across the board. because it permits efficient transportation of goods and services throughout the country's many regions. In actuality, infrastructure forms the basis of business investment.

Chapter 4

The distinction between social and economic infrastructure

What is Economic Infrastructure?

Energy, transportation, and communication infrastructure are all included in it. Promoting industries such as the manufacturing and trading of goods and services is crucial in light of this. Additionally, the distribution and production processes inside an economy are directly improved by economic infrastructure. The economic system is directly supported by this infrastructure, to put it simply.

What is Social Infrastructure?

It encompasses housing, healthcare, and education infrastructure. It means providing all services that improve the caliber of human capital. Though it accomplishes some social goals, social infrastructure has an indirect impact on the economy rather than a direct one.

The distinction between social and economic infrastructure

Basis	Economic Infrastructure	Social Infrastructure
Meaning	It encompasses the infrastructure related to communication, transportation, and energy.	It encompasses housing, healthcare, and education infrastructure.
Assistance for the economy	The system of economics is directly supported by economic infrastructure.	In a roundabout way, social infrastructure helps the economy.
Aids in	It enhances the economy's physical capital stock and contributes to the	It boosts the amount of human capital available to the economy and

	betterment of economic resources.	helps to improve the quality of human resources.
Example	Roads, trains, waterways, telecommunications, electricity, water supply, etc.	Sanitation, housing, education, health services, etc.

kinds of infrastructure

A few instances of different kinds of infrastructure are given below:

Business infrastructure

This comprises all of the buildings and infrastructure needed for a firm to start to run smoothly. These buildings and infrastructure might be crucial for a given industry or maybe for every industry in a given region. While communication services are necessary for all businesses to operate, a bakery's needs are usually different from those of a furniture manufacturer. To maintain profitability, businesses need infrastructure, and they work to accomplish their objectives by organizing all of the resources and instruments at their disposal.

IT Infrastructure

These are the non-physical and tangible parts that make using IT equipment and procedures easier. This infrastructure is used by communities, companies, and individuals to develop, provide, manage, and support IT-related services. The hardware, software, networks, and facilities that make up IT infrastructure are all necessary for the advancement and sustainability of contemporary communities. The effective deployment of IT infrastructures is becoming more and more crucial for nations and companies to stay sustainable and relevant in an increasingly technologically dependent world.

Ecological infrastructure

This is used to describe infrastructure and programs that offer natural answers to societal problems. This infrastructure could be the result of human involvement in certain cases and natural occurrence in others. Examples of this kind of infrastructure include wetlands, parks and gardens, athletic fields, coastal environments, pathways, and rivers. It is necessary because it preserves human life and the ecosystem. Green infrastructure serves environmental, social, and economic goals in addition to stormwater and climate change mitigation.

Robust infrastructure

This is a reference to constructions that can anticipate, stop, or lessen environmental disruptions. Engineers and builders create these kinds of structures intending to adapt to both predictable and unpredictable environmental events, including earthquakes, floods, tsunamis, and droughts. This kind of infrastructure is funded by local governments and communities in hurricane-prone areas. Companies can also create robust infrastructures that continue to function even in the face of environmental disasters.

Social infrastructure

This comprises the actual buildings and areas that facilitate the provision of fundamental social services including housing, healthcare, and education. Members of the community depend on these services for their health and financial security. Social housing, courts, hospitals, emergency services, libraries, recreational centers, and cultural organizations are a few instances of social infrastructure. These services promote social capital development, skill transfer, and the building of strong communities by aiding in the establishment and maintenance of connections.

Transport

People, services, and things may move around thanks to this infrastructure. It helps to make trade easier between domestic and foreign sites and has a big effect on a nation's economy. Because this kind of infrastructure frequently lasts a long time, investors find it appealing. Examples of this kind of infrastructure include pipelines, seaports, airports, highways, bridges, and railroads.

Power

Modern economies around the world are propelled forward by this infrastructure. It is crucial for many industries, as well as for the development of the economy, the generation of jobs, and growth. One form of energy is electricity, which encompasses its production, movement, and distribution. Transportation, refining, and oil and natural gas exploration are some of the additional elements. Energy supply to customers is made easier by the digitalization of some components of the energy infrastructure, such as intelligent building systems and smart meters.

Water

Because it supplies drinkable water, which is necessary for both health and economic growth, this infrastructure is vital to both life and industry. This pertains to the tangible infrastructures that guarantee the provision of drinkable water to homes and businesses, as well as the gathering, processing, and release of sewage. Water resource management, flood prevention, hydropower, and water-based transportation systems are also included. Water may play a crucial role in a nation's agricultural endeavors and electrical generation, provided that the right infrastructure is in place.

Chapter 5

Constructing routes to escape the poverty cycle through infrastructure

Taking Advantage of Public-Private Partnerships and Technology

The difficulties of constructing infrastructure in low-income nations must be tackled with creative ideas. Two important instruments that can be used to solve this issue are technology and public-private partnerships (PPPs). To complete infrastructure projects, the public and private sectors work together through PPPs. Even so, technology may enhance service delivery, boost productivity, and lower expenses. Creative ideas offer a chance to construct a framework for sustainable development and a way out of the poverty cycle.

Here are some detailed explanations of how to use PPPs and technology to develop infrastructure:

❖ **Joint strategy:** Combining the use of technology and public-private partnerships (PPPs) can significantly influence the development of infrastructure. For instance, in Ghana, rural areas are receiving solar-powered mini-grids from the government in collaboration with private businesses. Technology is being utilized to give a sustainable and affordable solution to the lack of energy in rural regions, while a PPP strategy is being employed to deliver the infrastructure as part of this project.

❖ **PPPs:** PPPs share the risks and rewards between the public and private sectors, which can assist overcome the absence of government support. Infrastructure projects can benefit from PPPs' innovative ideas, new technologies, and skilled labor. Furthermore, PPPs might assist in lessening the load on governments to enhance service delivery and supply infrastructure. As an illustration, the Mombasa-Nairobi Standard Gauge Railway in

Kenya was built through a public-private partnership (PPP). This project has increased the flow of people and products between the two cities by providing a state-of-the-art railway infrastructure.

The lack of infrastructure in low-income nations can be creatively addressed by utilizing PPPs and technology. Through the creation of possibilities for sustainable development, these projects can aid in the construction of pathways out of the poverty trap.

Chapter 6

The Contribution of the Private Sector to Bridging the Infrastructure Gap

Any country's ability to prosper economically depends on its infrastructure spending. Nonetheless, there is a huge infrastructure deficit because numerous governments worldwide are unable to sufficiently fund infrastructure initiatives. In developing nations, where a lack of infrastructure is a serious obstacle to economic expansion and the fight against poverty, this disparity is especially pronounced. Infrastructure investments from the private sector are crucial in addressing this problem. The provision of funds, knowledge, and resources by the private sector to finance and execute infrastructure projects holds great potential for mitigating the infrastructure deficit.

The following salient points underscore the significance of private-sector investments in narrowing the infrastructural disparity:

➢ **Employment creation:** Investing in infrastructure can boost employment and the economy. By generating jobs in the private sector, which can aid in the development of regional economies and encourage sustainable growth, private sector investments can enhance this effect even further. For instance, building projects brought about by investments from the private sector can support local economies by giving residents of the area work possibilities.

➢ **Risk-sharing:** Public and private sector investors may split the risks involved in infrastructure projects. This can ensure that private sector enterprises are driven to produce high-quality infrastructure projects that fulfill public demands, while simultaneously helping governments and taxpayers bear a smaller financial burden.

➢ **Knowledge:** The technical know-how and experience required to plan and carry out complicated infrastructure projects are frequently possessed by private sector businesses. Additionally, there are stronger incentives for private businesses to control risks, cut expenses, and optimize project designs. The private sector can also use innovation and technology to raise the caliber and productivity of infrastructure projects.

➢ **Finances:** One important source of finance for infrastructure projects might come from private sector investments. The private sector has access to capital markets and can borrow money through debt and equity financing, unlike governments, which have restricted resources. Big infrastructure projects like power plants, roadways, seaports, and airports become easier for private businesses to fund as a result.

To close the infrastructure deficit and spur economic growth, the private sector must invest heavily in infrastructure. Infrastructure development for poverty alleviation and economic growth can be facilitated by governments with the support of private sector investments, which offer resources, experience, and finance.

Chapter 7

The Function of Multilateral Organizations and Governments in Closing the Infrastructure Gap

In many regions of the world, poor infrastructure is a major barrier to growth and development. People may find it difficult to commute to work, receive basic services, and take advantage of economic progress if there is inadequate infrastructure, such as roads, bridges, and ports. One cannot emphasize how important governments and multilateral organizations are to closing the infrastructure gap. These organizations can strive to enhance infrastructure in some ways, from planning and financing to managing and maintaining it.

Information sharing and capacity building: To enhance infrastructure development, governments and international organizations can also contribute to information sharing and capacity building. Promoting innovation in infrastructure design and construction, exchanging technical know-how and best practices, and training local officials and engineers can all be part of this. To enhance local government's ability to design and carry out climate-resilient infrastructure projects, for instance, the Asian Development Bank's Urban Climate Change Resilience Trust Fund is in operation.

Planning and prioritization: Organizing and setting priorities for infrastructure projects is a crucial task for national and international organizations. This may entail determining the most important infrastructure requirements, working with other interested parties to coordinate, and creating a long-term plan for infrastructure development. To coordinate financing and implementation, the African Union's Program for Infrastructure Development in Africa (PIDA) has identified key infrastructure projects around the continent.

Providing funding: Funding infrastructure projects is one of the most important ways that national and international organizations may close the infrastructure gap. Grants, loans, and other financial aid are among the possible formats for this. To assist poor nations in building infrastructure, the World Bank's International Development Association (IDA) offers grants and low-interest loans.

Promoting development and lowering poverty depend heavily on the involvement of governments and multilateral organizations in closing the infrastructural gap. These organizations may guarantee that people throughout the world have access to the infrastructure they require to prosper by offering funding, organizing and prioritizing infrastructure projects, developing capacity, and exchanging knowledge.

Chapter 8

The Difference between Wealthy and Developing Nations

The infrastructure gap, or the difference between developed and developing nations, is a serious issue that has been plaguing developing nations for many years. Any nation's social and economic advancement depends heavily on its infrastructure. It alludes to the fundamental organizational and physical structures and infrastructure required for a civilization to function. Roads, bridges, ports, airports, trains, water supply, sanitary facilities, power, and telecommunications are some of these. Nonetheless, industrialized and developing nations' infrastructures differ greatly from one another. While emerging nations lack the fundamental infrastructure needed for their development, developed nations have well-established infrastructure.

The lack of infrastructure is a complicated problem with many underlying causes and effects on developing nations. Here are a few key understandings of this issue:

effects on development: The infrastructure divide significantly affects the growth of developing nations. People find it more difficult to access markets, healthcare, and education when there is poor quality infrastructure causing high transportation costs. Both social and economic development are slowed down as a result.

Lack of competent workers: Developing nations frequently struggle with a manpower shortage, which makes infrastructure planning and construction challenging. The cost of development is raised since many emerging nations must rely on outside knowledge to plan and construct their infrastructure.

Corruption: In many developing nations, corruption is a serious problem that has an impact on the growth of infrastructure. Money intended for infrastructure development is frequently diverted to other projects due to corruption, which results in resource mismanagement. This leads to sluggish development and low-quality infrastructure.

Political instability: The infrastructure deficit is also influenced by political instability. Potential investors are reluctant to fund infrastructure projects in nations with fragile political systems. Development is further slowed down as a result of a lack of investment.

Funding: The scarcity of funding is one of the main causes of the infrastructure deficit. The funds available to invest in the construction of infrastructure are frequently scarce in developing nations. Developing nations are unable to afford the significant costs associated with creating and maintaining infrastructure. Consequently, they become dependent on foreign assistance and loans from global institutions, potentially resulting in debt issues.

One major issue that hinders the growth of emerging nations is the lack of infrastructure. A multifaceted strategy including more funding, decreased corruption, political stability, and the development of trained workers is needed to address this issue. Developing nations can enhance their social and economic development as well as create escape routes from the poverty trap by addressing the infrastructure gap.

The Developing Countries' High Cost of Inadequate Infrastructure

Inadequate infrastructure is a problem that developing nations frequently deal with, as it feeds the poverty cycle. Transportation and healthcare are only two areas where a nation's economy can be severely impacted by inadequate infrastructure. Insufficient transportation infrastructure can lead to exorbitant transportation expenses, hence impeding firms' ability to convey goods and individuals' ability to obtain healthcare facilities. A nation's tourism sector may suffer from this lack of infrastructure, which would reduce traveler appeal. A lack of access to sanitary facilities and clean water can also result from inadequate infrastructure, which can lead to several health issues including waterborne infections.

To give comprehensive details regarding the enormous price that developing nations bear for having inadequate infrastructure, the following are important aspects to remember:

Effects on education: Inadequate facilities can hurt learning. For instance, children may find it challenging to learn and achieve if there is insufficient infrastructure in the school. Because fewer chances will be available to future generations, this could prolong the cycle of poverty.

Impact on the environment: The environment can be harmed by poor infrastructure. Pollution and environmental damage, for instance, can result from inadequate waste management infrastructure. The residents of the impacted areas may see considerable changes in their health and general well-being as a result.

Knowing How Poverty and Infrastructure Are Related

The foundation of any civilization is its infrastructure, which offers the support required to ensure that homes, companies, and the economy all run smoothly. While emerging nations find it difficult to supply the need for basic infrastructure, developed nations have well-established and maintained infrastructure. Poverty in developing nations is mostly caused by a lack of infrastructure. The poverty cycle is made worse by people's restricted access to basic services including healthcare, education, power, and water when there is insufficient infrastructure.

It is crucial to look at the several aspects that lead to a lack of infrastructure to comprehend the connection between poverty and it. Here are some observations that may aid in a deeper comprehension of the problem:

climate change: The absence of infrastructure is a result of climate change as well. Extreme weather conditions, such as droughts and floods, deteriorate already existing infrastructure and make it challenging to build new infrastructure.

In many developing nations, there are clear examples of the relationship between infrastructure and poverty. For example, 40% of people in sub-Saharan Africa do not have access to clean drinking water, and more than 60% of the population does not have access to electricity. People are unable to access contemporary technology and raise their standard of living without power. People who lack access to clean water may experience health problems, which may exacerbate their poverty.

Urbanization on the rise: The infrastructure that is now in place is under tremendous strain due to the fast urbanization of developing nations. Because of the cities' incapacity to meet the growing demand, the infrastructure and services are insufficient.

Bad governance: The development of infrastructure is hampered by political unrest and corruption. Infrastructure initiatives frequently result in unfinished or subparly built infrastructure because funds intended for them frequently wind up in the coffers of dishonest authorities.

Insufficient financial resources: Developing nations frequently lack the capital required to make infrastructure investments. Due to the perceived dangers and low returns, the private sector is reluctant to participate in infrastructure projects, and the government has a restricted budget.

It is indisputable that infrastructure and poverty are related. Inadequate infrastructure makes it more difficult to escape the vicious cycle of poverty and inhibits economic progress. To end the poverty trap, infrastructure development must be given top priority and funding, and governments, businesses, and the global community must collaborate.

Conclusion

Infrastructure of most types might be supplied privately. The efficiency with which any type of infrastructure is delivered will be influenced by incentives, regardless of whether it is provided by the government, the private sector, or privately with government subsidies. There would probably be fewer examples of government spending on inefficient infrastructure projects that displace private investment if user fees or private money were used instead. The intention is for the public and private sectors to collaborate to provide infrastructure for our communities.